AL Workbook Common Core Standards Edition

Published by
TOPICAL REVIEW BOOK COMPANY
P. O. Box 328
Onsted, MI 49265-0328
www.topicalrbc.com

EXAM	PAGE
Reference Sheet	i
June 2018	1
August 2018	11
January 2019	21
June 2019	32
August 2019	43
January 2020	53

THE STATE EDUCATION DEPARTMENT / THE UNIVERSITY OF THE STATE OF NEW YORK / ALBANY, NY 12234

Common Core High School Math Reference Sheet
(Algebra I, Geometry, Algebra II)

CONVERSIONS

1 inch = 2.54 centimeters	1 kilometer = 0.62 mile	1 cup = 8 fluid ounces
1 meter = 39.37 inches	1 pound = 16 ounces	1 pint = 2 cups
1 mile = 5280 feet	1 pound = 0.454 kilograms	1 quart = 2 pints
1 mile = 1760 yards	1 kilogram = 2.2 pounds	1 gallon = 4 quarts
1 mile = 1.609 kilometers	1 ton = 2000 pounds	1 gallon = 3.785 liters
		1 liter = 0.264 gallon
		1 liter = 1000 cubic centimeters

FORMULAS

Triangle	$A = \frac{1}{2}bh$	Pythagorean Theorem	$a^2 + b^2 = c^2$
Parallelogram	$A = bh$	Quadratic Formula	$x = \frac{-b \pm \sqrt{b^2 - 4ac}}{2a}$
Circle	$A = \pi r^2$	Arithmetic Sequence	$a_n = a_1 + (n-1)d$
Circle	$C = \pi d$ or $C = 2\pi r$	Geometric Sequence	$a_n = a_1 r^{n-1}$
General Prisms	$V = Bh$	Geometric Series	$S_n = \frac{a_1 - a_1 r^n}{1 - r}$ where $r \neq 1$
Cylinder	$V = \pi r^2 h$	Radians	1 radian = $\frac{180}{\pi}$ degrees
Sphere	$V = \frac{4}{3}\pi r^3$	Degrees	1 degree = $\frac{\pi}{180}$ radians
Cone	$V = \frac{1}{3}\pi r^2 h$	Exponential Growth/Decay	$A = A_0 e^{k(t - t_0)} + B_0$
Pyramid	$V = \frac{1}{3}Bh$		

ALGEBRA 2 – COMMON CORE
June 2018
Part I

Answer all 24 questions in this part. Each correct answer will receive 2 credits. No partial credit will be allowed. Utilize the information provided for each question to determine your answer. Note that diagrams are not necessarily drawn to scale. For each statement or question, choose the word or expression that, of those given, best completes the statement or answers the question. Record your answers in the space provided. [48]

1. The graphs of the equations $y = x^2 + 4x - 1$ and $y + 3 = x$ are drawn on the same set of axes. One solution of this system is
(1) $(-5, -2)$ (2) $(-1, -4)$ (3) $(1, 4)$ (4) $(-2, -1)$

2. Which statement is true about the graph of $f(x) = \left(\frac{1}{8}\right)^x$?
(1) The graph is always increasing.
(2) The graph is always decreasing.
(3) The graph passes through $(1, 0)$.
(4) The graph has an asymptote, $x = 0$.

3. For all values of x for which the expression is defined, $\frac{x^3 + 2x^2 - 9x - 18}{x^3 - x^2 - 6x}$, in simplest form, is equivalent to
(1) 3 (2) $-\frac{17}{2}$ (3) $\frac{x+3}{x}$ (4) $\frac{x^2 - 9}{x(x-3)}$

4. A scatterplot showing the weight, w, in grams, of each crystal after growing t hours is shown. The relationship between weight, w, and time, t, is best modeled by
(1) $w = 4^t + 5$
(2) $w = (1.4)^t + 2$
(3) $w = 5(2.1)^t$
(4) $w = 8(.75)^t$

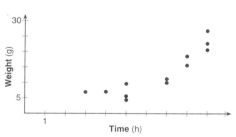

5. Where i is the imaginary unit, the expression $(x + 3i)^2 - (2x - 3i)^2$ is equivalent to
(1) $-3x^2$ (2) $-3x^2 - 18$ (3) $-3x^2 + 18xi$ (4) $-3x^2 - 6xi - 18$

6. Which function is even?
(1) $f(x) = \sin x$
(2) $f(x) = x^2 - 4$
(3) $f(x) = |x - 2| + 5$
(4) $f(x) = x^4 + 3x^3 + 4$

7. The function $N(t) = 100e^{-0.023t}$ models the number of grams in a sample of cesium-137 that remain after t years. On which interval is the sample's average rate of decay the fastest?
(1) $[1, 10]$ (2) $[10, 20]$ (3) $[15, 25]$ (4) $[1, 30]$

ALGEBRA 2 – COMMON CORE
June 2018

8. Which expression can be rewritten as $(x+7)(x-1)$?

(1) $(x+3)^2 - 16$

(2) $(x+3)^2 - 10(x+3) - 2(x+3) + 20$

(3) $\dfrac{(x-1)(x^2-6x-7)}{(x+1)}$

(4) $\dfrac{(x+7)(x^2+4x+3)}{(x+3)}$

8 _____

9. What is the solution set of the equation $\dfrac{2}{x} - \dfrac{3x}{x+3} = \dfrac{x}{x+3}$?

(1) $\{3\}$ (2) $\left\{\dfrac{3}{2}\right\}$ (3) $\{-2, 3\}$ (4) $\left\{-1, \dfrac{3}{2}\right\}$

9 _____

10. The depth of the water at a marker 20 feet from the shore in a bay is depicted in the graph.

If the depth, d, is measured in feet and time, t, is measured in hours since midnight, what is an equation for the depth of the water at the marker?

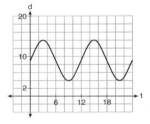

(1) $d = 5\cos\left(\dfrac{\pi}{6}t\right) + 9$ (3) $d = 9\sin\left(\dfrac{\pi}{6}t\right) + 5$

(2) $d = 9\cos\left(\dfrac{\pi}{6}t\right) + 5$ (4) $d = 5\sin\left(\dfrac{\pi}{6}t\right) + 9$

10 _____

11. On a given school day, the probability that Nick oversleeps is 48% and the probability he has a pop quiz is 25%. Assuming these two events are independent, what is the probability that Nick oversleeps and has a pop quiz on the same day?
(1) 73%   (2) 36%   (3) 23%   (4) 12%

11 _____

12. If $x-1$ is a factor of $x^3 - kx^2 + 2x$, what is the value of k?
(1) 0   (2) 2   (3) 3   (4) –3

12 _____

13. The profit function, $p(x)$, for a company is the cost function, $c(x)$, subtracted from the revenue function, $r(x)$. The profit function for the Acme Corporation is $p(x) = -0.5x^2 + 250x - 300$ and the revenue function is $r(x) = -0.3x^2 + 150x$. The cost function for the Acme Corporation is
(1) $c(x) = 0.2x^2 - 100x + 300$ (3) $c(x) = -0.2x^2 + 100x - 300$
(2) $c(x) = 0.2x^2 + 100x + 300$ (4) $c(x) = -0.8x^2 + 400x - 300$

13 _____

14. The populations of two small towns at the beginning of 2018 and their annual population growth rate are shown in the table.

Town	Population	Annual Population Growth Rate
Jonesville	1240	6% increase
Williamstown	890	11% increase

Assuming the trend continues, approximately how many years after the beginning of 2018 will it take for the populations to be equal?
(1) 7   (2) 20   (3) 68   (4) 125

14 _____

ALGEBRA 2 – COMMON CORE
June 2018

15. What is the inverse of $f(x) = x^3 - 2$?
(1) $f^{-1}(x) = \sqrt[3]{x} + 2$
(2) $f^{-1}(x) = \pm\sqrt[3]{x} + 2$
(3) $f^{-1}(x) = \sqrt[3]{x+2}$
(4) $f^{-1}(x) = \pm\sqrt[3]{x+2}$

15 _____

16. A 4th degree polynomial has zeros –5, 3, i, and –i. Which graph could represent the function defined by this polynomial?

  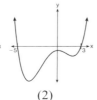 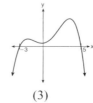 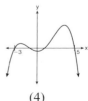

(1)   (2)   (3)   (4)

16 _____

17. The weights of bags of Graseck's Chocolate Candies are normally distributed with a mean of 4.3 ounces and a standard deviation of 0.05 ounces. What is the probability that a bag of these chocolate candies weighs less than 4.27 ounces?
(1) 0.2257   (2) 0.2743   (3) 0.7257   (4) 0.7757

17 _____

18. The half-life of iodine-131 is 8 days. The percent of the isotope left in the body d days after being introduced is $I = 100\left(\frac{1}{2}\right)^{\frac{d}{8}}$. When this equation is written in terms of the number e, the base of the natural logarithm, it is equivalent to $I = 100e^{kd}$. What is the approximate value of the constant, k?
(1) –0.087   (2) 0.087   (3) –11.542   (4) 11.542

18 _____

19. The graph of $y = \log_7 x$ is translated to the right 1 unit and down 1 unit. The coordinates of the x-intercept of the translated graph are
(1) (0, 0)   (2) (1, 0)   (3) (2, 0)   (4) (3, 0)

19 _____

20. For positive values of x, which expression is equivalent to $\sqrt{16x^2} \bullet x^{\frac{2}{3}} + \sqrt[3]{8x^5}$?
(1) $6\sqrt[5]{x^3}$ (2) $6\sqrt[3]{x^5}$ (3) $4\sqrt[3]{x^2} + 2\sqrt[3]{x^5}$ (4) $4\sqrt{x^3} + 2\sqrt[5]{x^3}$

20 _____

21. Which equation represents a parabola with a focus of (–2, 5) and a directrix of $y = 9$?
(1) $(y - 7)^2 = 8(x + 2)$
(2) $(y - 7)^2 = -8(x + 2)$
(3) $(x + 2)^2 = 8(y - 7)$
(4) $(x + 2)^2 = -8(y - 7)$

21 _____

22. Given the following polynomials

$x = (a + b + c)^2$
$y = a^2 + b^2 + c^2$
$z = ab + bc + ac$

Which identity is true?
(1) $x = y - z$ (2) $x = y + z$ (3) $x = y - 2z$ (4) $x = y + 2z$

22 _____

23. On average, college seniors graduating in 2012 could compute their growing student loan debt using the function $D(t) = 29{,}400(1.068)^t$, where t is time in years. Which expression is equivalent to $29{,}400(1.068)^t$ and could be used by students to identify an approximate daily interest rate on their loans?

(1) $29{,}400\left(1.068^{\frac{1}{365}}\right)^t$

(2) $29{,}400\left(\dfrac{1.068}{365}\right)^{365t}$

(3) $29{,}400\left(1 + \dfrac{0.068}{365}\right)^t$

(4) $29{,}400\left(1.068^{\frac{1}{365}}\right)^{365t}$

23 _____

24. A manufacturing plant produces two different-sized containers of peanuts. One container weighs x ounces and the other weighs y pounds. If a gift set can hold one of each size container, which expression represents the number of gift sets needed to hold 124 ounces?

(1) $\dfrac{124}{16x + y}$ (2) $\dfrac{x + 16y}{124}$ (3) $\dfrac{124}{x + 16y}$ (4) $\dfrac{16x + y}{124}$

24 _____

Part II

Answer all 8 questions in this part. Each correct answer will receive 2 credits. Clearly indicate the necessary steps, including appropriate formula substitutions, diagrams, graphs, charts, etc. Utilize the information provided for each question to determine your answer. Note that diagrams are not necessarily drawn to scale. For all questions in this part, a correct numerical answer with no work shown will receive only 1 credit. All answers should be written in pen, except for graphs and drawings, which should be done in pencil. [16]

25. A survey about television-viewing preferences was given to randomly selected freshmen and seniors at Fairport High School. The results are shown in the table.

Favorite Type of Program

	Sports	Reality Show	Comedy Series
Senior	83	110	67
Freshman	119	103	54

A student response is selected at random from the results. State the *exact* probability the student response is from a freshman, given the student prefers to watch reality shows on television.

26. On the grid below, graph the function $f(x) = x^3 - 6x^2 + 9x + 6$ on the domain $-1 \leq x \leq 4$.

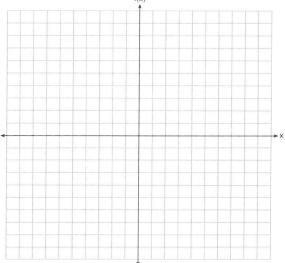

27. Solve the equation $2x^2 + 5x + 8 = 0$. Express the answer in $a + bi$ form.

28. Chuck's Trucking Company has decided to initiate an Employee of the Month program. To determine the recipient, they put the following sign on the back of each truck.

> How's My Driving?
> Call 1-555-DRIVING

The driver who receives the highest number of positive comments will win the recognition. Explain *one* statistical bias in this data collection method.

29. Determine the quotient and remainder when $(6a^3 + 11a^2 - 4a - 9)$ is divided by $(3a - 2)$. Express your answer in the form $q(a) + \dfrac{r(a)}{d(a)}$.

30. The recursive formula to describe a sequence is shown.
$a_1 = 3$
$a_n = 1 + 2a_{n-1}$

State the first four terms of this sequence.

Can this sequence be represented using an explicit geometric formula? Justify your answer.

31. The Wells family is looking to purchase a home in a suburb of Rochester with a 30-year mortgage that has an annual interest rate of 3.6%. The house the family wants to purchase is $152,500 and they will make a $15,250 down payment and borrow the remainder. Use the formula below to determine their monthly payment, to the *nearest dollar*.

$$M = \dfrac{P\left(\dfrac{r}{12}\right)\left(1+\dfrac{r}{12}\right)^n}{\left(1+\dfrac{r}{12}\right)^n - 1}$$

M = monthly payment
P = amount borrowed
r = annual interest rate
n = total number of monthly payments

32. An angle, θ, is in standard position and its terminal side passes through the point $(2, -1)$. Find the *exact* value of sin θ.

Part III

Answer all 4 questions in this part. Each correct answer will receive 4 credits. Clearly indicate the necessary steps, including appropriate formula substitutions, diagrams, graphs, charts, etc. Utilize the information provided for each question to determine your answer. Note that diagrams are not necessarily drawn to scale. For all questions in this part, a correct numerical answer with no work shown will receive only 1 credit. All answers should be written in pen, except for graphs and drawings, which should be done in pencil. [16]

33. Solve algebraically for all values of x: $\sqrt{6-2x} + x = 2(x+15) - 9$

34. Joseph was curious to determine if scent improves memory. A test was created where better memory is indicated by higher test scores. A controlled experiment was performed where one group was given the test on scented paper and the other group was given the test on unscented paper. The summary statistics from the experiment are given below.

Calculate the difference in means in the experimental test grades (scented − unscented).

	Scented Paper	Unscented Paper
$\bar{x}$	23	18
s_x	2.898	2.408

A simulation was conducted in which the subjects' scores were rerandomized into two groups 1000 times. The differences of the group means were calculated each time. The results are shown below.

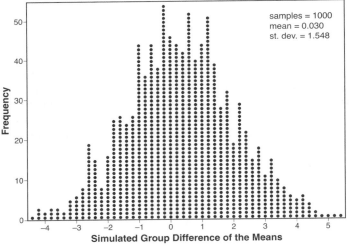

Use the simulation results to determine the interval representing the middle 95% of the difference in means, to the *nearest hundredth*.

Is the difference in means in Joseph's experiment statistically significant based on the simulation? Explain.

35. Carla wants to start a college fund for her daughter Lila. She puts $63,000 into an account that grows at a rate of 2.55% per year, compounded monthly. Write a function, $C(t)$, that represents the amount of money in the account t years after the account is opened, given that no more money is deposited into or withdrawn from the account.

Calculate algebraically the number of years it will take for the account to reach $100,000, to the *nearest hundredth of a year*.

36. The height, $h(t)$ in cm, of a piston, is given by the equation $h(t) = 12\cos(\frac{\pi}{3}t) + 8$, where t represents the number of seconds since the measurements began.

Determine the average rate of change, in cm/sec, of the piston's height on the interval $1 \le t \le 2$.

At what value(s) of t, to the *nearest tenth of a second*, does $h(t) = 0$ in the interval $1 \le t \le 5$? Justify your answer.

Part IV

Answer the question in this part. A correct answer will receive 6 credits. Clearly indicate the necessary steps, including appropriate formula substitutions, diagrams, graphs, charts, etc. Utilize the information provided to determine your answer. Note that diagrams are not necessarily drawn to scale. A correct numerical answer with no work shown will receive only 1 credit. All answers should be written in pen, except for graphs and drawings, which should be done in pencil. [6]

37. Website popularity ratings are often determined using models that incorporate the number of visits per week a website receives. One model for ranking websites is $P(x) = \log(x - 4)$, where x is the number of visits per week in thousands and $P(x)$ is the website's popularity rating. According to this model, if a website is visited 16,000 times in one week, what is its popularity rating, rounded to the *nearest tenth*?

Graph $y = P(x)$ on the axes below.

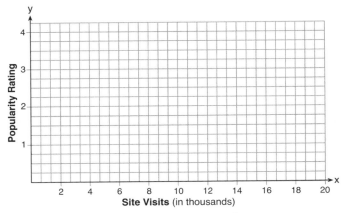

An alternative rating model is represented by $R(x) = \frac{1}{2}x - 6$, where x is the number of visits per week in thousands. Graph $R(x)$ on the same set of axes. For what number of weekly visits will the two models provide the same rating?

ALGEBRA 2 – COMMON CORE

August 2018

Part I

Answer all 24 questions in this part. Each correct answer will receive 2 credits. No partial credit will be allowed. Utilize the information provided for each question to determine your answer. Note that diagrams are not necessarily drawn to scale. For each statement or question, choose the word or expression that, of those given, best completes the statement or answers the question. Record your answers in the space provided. [48]

1. The solution of $87e^{0.3x} = 5918$, to the *nearest thousandth*, is
(1) 0.583　　(2) 1.945　　(3) 4.220　　(4) 14.066　　　1 _____

2. A researcher randomly divides 50 bean plants into two groups. He puts one group by a window to receive natural light and the second group under artificial light. He records the growth of the plants weekly. Which data collection method is described in this situation?
(1) observational study　　(3) survey
(2) controlled experiment　　(4) systematic sample　　　2 _____

3. If $f(x) = x^2 + 9$ and $g(x) = x + 3$, which operation would *not* result in a polynomial expression?
(1) $f(x) + g(x)$　　(2) $f(x) - g(x)$　　(3) $f(x) \cdot g(x)$　　(4) $f(x) \div g(x)$　　　3 _____

4. Consider the function $p(x) = 3x^3 + x^2 - 5x$ and the graph of $y = m(x)$.

Which statement is true?

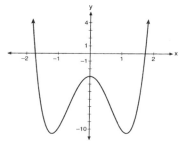

(1) $p(x)$ has three real roots and $m(x)$ has two real roots.
(2) $p(x)$ has one real root and $m(x)$ has two real roots.
(3) $p(x)$ has two real roots and $m(x)$ has three real roots.
(4) $p(x)$ has three real roots and $m(x)$ has four real roots.　　　4 _____

5. Which expression is equivalent to $\dfrac{2x^4 + 8x^3 - 25x^2 - 6x + 14}{x+6}$?
(1) $2x^3 + 4x^2 + x - 12 + \dfrac{86}{x+6}$　　(3) $2x^3 - 4x^2 - x + \dfrac{14}{x+6}$
(2) $2x^3 - 4x^2 - x + 14$　　(4) $2x^3 - 4x^2 - x$　　　5 _____

6. Given $f(x) = \dfrac{1}{2}x + 8$, which equation represents the inverse, $g(x)$?
(1) $g(x) = 2x - 8$　　(3) $g(x) = -\dfrac{1}{2}x + 8$
(2) $g(x) = 2x - 16$　　(4) $g(x) = -\dfrac{1}{2}x - 16$　　　6 _____

ALGEBRA 2 – COMMON CORE
August 2018

7. The value(s) of x that satisfy $\sqrt{x^2 - 4x - 5} = 2x - 10$ are
(1) {5}   (2) {7}   (3) {5, 7}   (4) {3, 5, 7}   7 _____

8. Stephanie found that the number of white-winged crossbills in an area can be represented by the formula $C = 550(1.08)^t$, where t represents the number of years since 2010. Which equation correctly represents the number of white-winged crossbills in terms of the monthly rate of population growth?
(1) $C = 550(1.00643)^t$
(2) $C = 550(1.00643)^{12t}$
(3) $C = 550(1.00643)^{\frac{t}{12}}$
(4) $C = 550(1.00643)^{t+12}$ 8 _____

9. The roots of the equation $3x^2 + 2x = -7$ are
(1) $-2, -\dfrac{1}{3}$ (2) $-\dfrac{7}{3}, 1$ (3) $-\dfrac{1}{3} \pm \dfrac{2i\sqrt{5}}{3}$ (4) $-\dfrac{1}{3} \pm \dfrac{\sqrt{11}}{3}$ 9 _____

10. The average depreciation rate of a new boat is approximately 8% per year. If a new boat is purchased at a price of $75,000, which model is a recursive formula representing the value of the boat n years after it was purchased?
(1) $a_n = 75{,}000(0.08)^n$
(2) $a_0 = 75{,}000$
    $a_n = (0.92)^n$
(3) $a_n = 75{,}000(1.08)^n$
(4) $a_0 = 75{,}000$
    $a_n = 0.92(a_{n-1})$ 10 _____

11. Given $\cos \theta = \dfrac{7}{25}$, where θ is an angle in standard position terminating in quadrant IV, and $\sin^2\theta + \cos^2\theta = 1$, what is the value of $\tan \theta$?
(1) $-\dfrac{24}{25}$ (2) $-\dfrac{24}{7}$ (3) $\dfrac{24}{25}$ (4) $\dfrac{24}{7}$ 11 _____

12. For $x > 0$, which expression is equivalent to $\dfrac{\sqrt[3]{x^2} \cdot \sqrt{x^5}}{\sqrt[6]{x}}$?
(1) x (2) $x^{\frac{3}{2}}$ (3) x^3 (4) x^{10} 12 _____

13. Jake wants to buy a car and hopes to save at least $5000 for a down payment. The table below summarizes the amount of money he plans to save each week.

Week	1	2	3	4	5
Money Saved, in Dollars	2	5	12.5	31.25	...

Based on this plan, which expression should he use to determine how much he has saved in n weeks?
(1) $\dfrac{2 - 2(2.5^n)}{1 - 2.5}$ (2) $\dfrac{2 - 2(2.5^{n-1})}{1 - 2.5}$ (3) $\dfrac{1 - 2.5^n}{1 - 2.5}$ (4) $\dfrac{1 - 2.5^{n-1}}{1 - 2.5}$ 13 _____

14. Which expression is equivalent to $x^6y^4(x^4 - 16) - 9(x^4 - 16)$?
(1) $x^{10}y^4 - 16x^6y^4 - 9x^4 - 144$
(2) $(x^6y^4 - 9)(x + 2)^3(x - 2)$
(3) $(x^3y^2 + 3)(x^3y^2 - 3)(x + 2)^2(x - 2)^2$
(4) $(x^3y^2 + 3)(x^3y^2 - 3)(x^2 + 4)(x^2 - 4)$ 14 _____

15. If $A = -3 + 5i$, $B = 4 - 2i$, and $C = 1 + 6i$, where i is the imaginary unit, then $A - BC$ equals
(1) $5 - 17i$ (2) $5 + 27i$ (3) $-19 - 17i$ (4) $-19 + 27i$

16. Which sketch best represents the graph of $x = 3^y$?

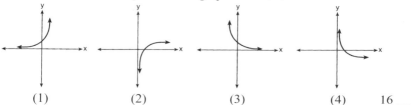

(1) (2) (3) (4)

17. The graph represents national and New York State average gas prices.

If New York State's gas prices are modeled by $G(x)$ and $C > 0$, which expression best approximates the national average x months from August 2014?

(1) $G(x + C)$
(2) $G(x) + C$
(3) $G(x - C)$
(4) $G(x) - C$

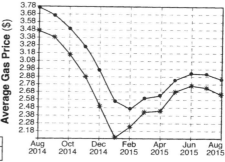

18. Data for the students enrolled in a local high school are shown in the Venn diagram. If a student from the high school is selected at random, what is the probability that the student is a sophomore given that the student is enrolled in Algebra II?

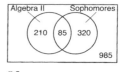

(1) $\dfrac{85}{210}$ (2) $\dfrac{85}{295}$ (3) $\dfrac{85}{405}$ (4) $\dfrac{85}{1600}$

19. If $p(x) = 2\ln(x) - 1$ and $m(x) = \ln(x + 6)$, then what is the solution for $p(x) = m(x)$?
(1) 1.65 (2) 3.14 (3) 5.62 (4) no solution

20. Which function's graph has a period of 8 and reaches a maximum height of 1 if at least one full period is graphed?
(1) $y = -4\cos\left(\dfrac{\pi}{4}x\right) - 3$ (3) $y = -4\cos(8x) - 3$
(2) $y = -4\cos\left(\dfrac{\pi}{4}x\right) + 5$ (4) $y = -4\cos(8x) + 5$

21. Given $c(m) = m^3 - 2m^2 + 4m - 8$, the solution of $c(m) = 0$ is
(1) ± 2 (2) 2, only (3) $2i$, 2 (4) $\pm 2i$, 2

22. The height above ground for a person riding a Ferris wheel after t seconds is modeled by $h(t) = 150\sin\left(\frac{\pi}{45}t + 67.5\right) + 160$ feet. How many seconds does it take to go from the bottom of the wheel to the top of the wheel?
(1) 10   (2) 45   (3) 90   (4) 150

23. The parabola described by the equation $y = \frac{1}{12}(x - 2)^2 + 2$ has the directrix at $y = -1$. The focus of the parabola is
(1) (2, −1)   (2) (2, 2)   (3) (2, 3)   (4) (2, 5)

24. A fast-food restaurant analyzes data to better serve its customers. After its analysis, it discovers that the events D, that a customer uses the drive-thru, and F, that a customer orders French fries, are independent. The following data are given in a report:

$$P(F) = 0.8$$
$$P(F \cap D) = 0.456$$

Given this information, $P(F|D)$ is
(1) 0.344   (2) 0.3648   (3) 0.57   (4) 0.8

Part II

Answer all 8 questions in this part. Each correct answer will receive 2 credits. Clearly indicate the necessary steps, including appropriate formula substitutions, diagrams, graphs, charts, etc. Utilize the information provided for each question to determine your answer. Note that diagrams are not necessarily drawn to scale. For all questions in this part, a correct numerical answer with no work shown will receive only 1 credit. All answers should be written in pen, except for graphs and drawings, which should be done in pencil. [16]

25. Over the set of integers, factor the expression $x^4 - 4x^2 - 12$.

26. Express the fraction $\dfrac{2x^{\frac{3}{2}}}{\left(16^4\right)^{\frac{1}{4}}}$ in simplest radical form.

27. The world population was 2560 million people in 1950 and 3040 million in 1960 and can be modeled by the function $p(t) = 2560e^{0.017185t}$, where t is time in years after 1950 and $p(t)$ is the population in millions. Determine the average rate of change of $p(t)$ in *millions of people per year*, from $4 \leq t \leq 8$. Round your answer to the *nearest hundredth*.

28. The scores of a recent test taken by 1200 students had an approximately normal distribution with a mean of 225 and a standard deviation of 18. Determine the number of students who scored between 200 and 245.

29. Algebraically solve for x: $\dfrac{-3}{x+3} + \dfrac{1}{2} = \dfrac{x}{6} - \dfrac{1}{2}$

30. Graph $t(x) = 3\sin(2x) + 2$ over the domain $[0, 2\pi]$ on the set of axes below.

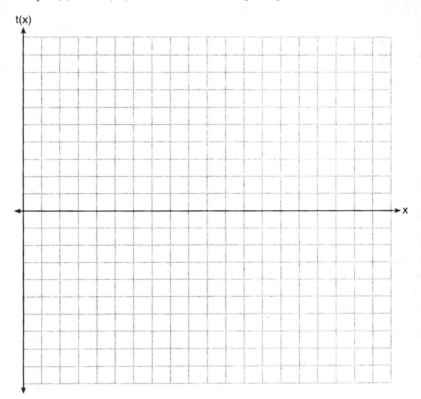

31. Solve the following system of equations algebraically.
$$x^2 + y^2 = 400$$
$$y = x - 28$$

32. Some smart-phone applications contain "in-app" purchases, which allow users to purchase special content within the application. A random sample of 140 users found that 35 percent made in-app purchases. A simulation was conducted with 200 samples of 140 users assuming 35 percent of the samples make in-app purchases. The approximately normal results are shown below.

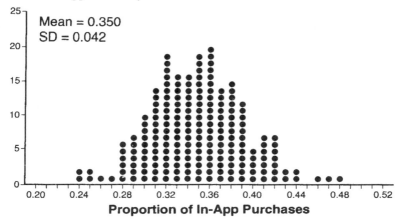

Considering the middle 95% of the data, determine the margin of error, to the *nearest hundredth*, for the simulated results. In the given context, explain what this value represents.

ALGEBRA 2 – COMMON CORE
August 2018
Part III

Answer all 4 questions in this part. Each correct answer will receive 4 credits. Clearly indicate the necessary steps, including appropriate formula substitutions, diagrams, graphs, charts, etc. Utilize the information provided for each question to determine your answer. Note that diagrams are not necessarily drawn to scale. For all questions in this part, a correct numerical answer with no work shown will receive only 1 credit. All answers should be written in pen, except for graphs and drawings, which should be done in pencil. [16]

33. Solve the following system of equations algebraically for all values of x, y, and z.

$$2x + 3y - 4z = -1$$
$$x - 2y + 5z = 3$$
$$-4x + y + z = 16$$

34. Evaluate $j(-1)$ given $j(x) = 2x^4 - x^3 - 35x^2 + 16x + 48$. Explain what your answer tells you about $x + 1$ as a factor.

Algebraically find the remaining zeros of $j(x)$.

35. Determine, to the *nearest tenth of a year*, how long it would take an investment to double at a $3\frac{3}{4}$% interest rate, compounded continuously.

36. To determine if the type of music played while taking a quiz has a relationship to results, 16 students were randomly assigned to either a room softly playing classical music or a room softly playing rap music. The results on the quiz were as follows:

Classical: 74, 83, 77, 77, 84, 82, 90, 89
Rap: 77, 80, 78, 74, 69, 72, 78, 69

John correctly rounded the difference of the means of his experimental groups as 7. How did John obtain this value and what does it represent in the given context? Justify your answer.

To determine if there is any significance in this value, John rerandomized the 16 scores into two groups of 8, calculated the difference of the means, and simulated this process 250 times as shown.

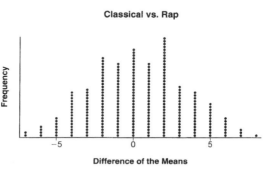

Does the simulation support the theory that there may be a significant difference in quiz scores? Explain.

ALGEBRA 2 – COMMON CORE
August 2018
Part IV

Answer the question in this part. A correct answer will receive 6 credits. Clearly indicate the necessary steps, including appropriate formula substitutions, diagrams, graphs, charts, etc. Utilize the information provided to determine your answer. Note that diagrams are not necessarily drawn to scale. A correct numerical answer with no work shown will receive only 1 credit. All answers should be written in pen, except for graphs and drawings, which should be done in pencil. [6]

37. A major car company analyzes its revenue, $R(x)$, and costs $C(x)$, in millions of dollars over a fifteen-year period. The company represents its revenue and costs as a function of time, in years, x, using the given functions.

$$R(x) = 550x^3 - 12{,}000x^2 + 83{,}000x + 7000$$
$$C(x) = 880x^3 - 21{,}000x^2 + 150{,}000x - 160{,}000$$

The company's profits can be represented as the difference between its revenue and costs. Write the profit function, $P(x)$, as a polynomial in standard form.

Graph $y = P(x)$ on the set of axes over the domain $2 \leq x \leq 16$.

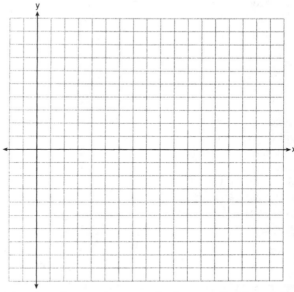

Over the given domain, state when the company was the least profitable and the most profitable, to the *nearest year*. Explain how you determined your answer.

ALGEBRA 2 – COMMON CORE
January 2019
Part I

Answer all 24 questions in this part. Each correct answer will receive 2 credits. No partial credit will be allowed. Utilize the information provided for each question to determine your answer. Note that diagrams are not necessarily drawn to scale. For each statement or question, choose the word or expression that, of those given, best completes the statement or answers the question. Record your answers in the space provided. [48]

1. Suppose two sets of test scores have the same mean, but different standard deviations, σ_1 and σ_2, with $\sigma_2 > \sigma_1$. Which statement best describes the variability of these data sets?
(1) Data set one has the greater variability.
(2) Data set two has the greater variability.
(3) The variability will be the same for each data set.
(4) No conclusion can be made regarding the variability of either set.

2. If $f(x) = \log_3 x$ and $g(x)$ is the image of $f(x)$ after a translation five units to the left, which equation represents $g(x)$?
(1) $g(x) = \log_3 (x + 5)$
(2) $g(x) = \log_3 x + 5$
(3) $g(x) = \log_3 (x - 5)$
(4) $g(x) = \log_3 x - 5$

3. When factoring to reveal the roots of the equation $x^3 + 2x^2 - 9x - 18 = 0$, which equations can be used?
    I. $x^2(x + 2) - 9(x + 2) = 0$
    II. $x(x^2 - 9) + 2(x^2 - 9) = 0$
    III. $(x - 2)(x^2 - 9) = 0$
(1) I and II, only   (2) I and III, only   (3) II and III, only   (4) I, II, and III

4. When a ball bounces, the heights of consecutive bounces form a geometric sequence. The height of the first bounce is 121 centimeters and the height of the third bounce is 64 centimeters. To the *nearest centimeter*, what is the height of the fifth bounce?
(1) 25   (2) 34   (3) 36   (4) 42

5. The solutions to the equation $5x^2 - 2x + 13 = 9$ are
(1) $\frac{1}{5} \pm \frac{\sqrt{21}}{5}$ (2) $\frac{1}{5} \pm \frac{\sqrt{19}}{5}i$ (3) $\frac{1}{5} \pm \frac{\sqrt{66}}{5}i$ (4) $\frac{1}{5} \pm \frac{\sqrt{66}}{5}$

6. Julia deposits $2000 into a savings account that earns 4% interest per year. The exponential function that models this savings account is $y = 2000(1.04)^t$, where t is the time in years. Which equation correctly represents the amount of money in her savings account in terms of the monthly growth rate?
(1) $y = 166.67(1.04)^{0.12t}$
(2) $y = 2000(1.01)^t$
(3) $y = 2000(1.0032737)^{12t}$
(4) $y = 166.67(1.0032737)^t$

7. Tides are a periodic rise and fall of ocean water. On a typical day at a seaport, to predict the time of the next high tide, the most important value to have would be the
(1) time between consecutive low tides
(2) time when the tide height is 20 feet
(3) average depth of water over a 24-hour period
(4) difference between the water heights at low and high tide

8. An estimate of the number of milligrams of a medication in the bloodstream t hours after 400 mg has been taken can be modeled by the function below.
$$I(t) = 0.5t^4 + 3.45t^3 - 96.65t^2 + 347.7t, \text{ where } 0 \leq t \leq 6$$
Over what time interval does the amount of medication in the bloodstream strictly increase?
(1) 0 to 2 hours   (2) 0 to 3 hours   (3) 2 to 6 hours   (4) 3 to 6 hours

9. Which representation of a quadratic has imaginary roots?

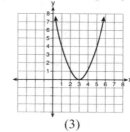

x	y
−2.5	2
−2.0	0
−1.5	−1
−1.0	−1
−0.5	0
0.0	2

(1)   (3)

$2(x + 3)^2 = 64$ $2x^2 + 32 = 0$
(2)   (4)

10. A random sample of 100 people that would best estimate the proportion of all registered voters in a district who support improvements to the high school football field should be drawn from registered voters in the district at a
(1) football game   (3) school fund-raiser
(2) supermarket   (4) high school band concert

11. Which expression is equivalent to $(2x - i)^2 - (2x - i)(2x + 3i)$ where i is the imaginary unit and x is a real number?
(1) $-4 - 8xi$ (2) $-4 - 4xi$ (3) 2   (4) $8x - 4i$

12. Suppose events A and B are independent and $P(A$ and $B)$ is 0.2. Which statement could be true?
(1) $P(A) = 0.4, P(B) = 0.3, P(A$ or $B) = 0.5$
(2) $P(A) = 0.8, P(B) = 0.25$
(3) $P(A|B) = 0.2, P(B) = 0.2$
(4) $P(A) = 0.15, P(B) = 0.05$

13. The function $f(x) = a \cos bx + c$ is plotted on the graph shown.

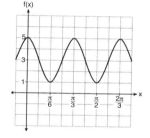

What are the values of a, b, and c?
(1) $a = 2, b = 6, c = 3$
(2) $a = 2, b = 3, c = 1$
(3) $a = 4, b = 6, c = 5$
(4) $a = 4, b = \frac{\pi}{3}, c = 3$

13 _____

14. Which equation represents the equation of the parabola with focus $(-3, 3)$ and directrix $y = 7$?
(1) $y = \frac{1}{8}(x + 3)^2 - 5$
(2) $y = \frac{1}{8}(x - 3)^2 + 5$
(3) $y = -\frac{1}{8}(x + 3)^2 + 5$
(4) $y = -\frac{1}{8}(x - 3)^2 + 5$

14 _____

15. What is the solution set of the equation $\frac{2}{3x+1} = \frac{1}{x} - \frac{6x}{3x+1}$?
(1) $\left\{-\frac{1}{3}, \frac{1}{2}\right\}$ (2) $\left\{-\frac{1}{3}\right\}$ (3) $\left\{\frac{1}{2}\right\}$ (4) $\left\{\frac{1}{3}, -2\right\}$

15 _____

16. Savannah just got contact lenses. Her doctor said she can wear them 2 hours the first day, and can then increase the length of time by 30 minutes each day. If this pattern continues, which formula would *not* be appropriate to determine the length of time, in either minutes or hours, she could wear her contact lenses on the nth day?
(1) $a_1 = 120$
   $a_n = a_{n-1} + 30$
(2) $a_n = 90 + 30n$
(3) $a_1 = 2$
   $a_n = a_{n-1} + 0.5$
(4) $a_n = 2.5 + 0.5n$

16 _____

17. If $f(x) = a^x$ where $a > 1$, then the inverse of the function is
(1) $f^{-1}(x) = \log_x a$
(2) $f^{-1}(x) = a \log x$
(3) $f^{-1}(x) = \log_a x$
(4) $f^{-1}(x) = x \log a$

17 _____

18. Kelly-Ann has $20,000 to invest. She puts half of the money into an account that grows at an annual rate of 0.9% compounded monthly. At the same time, she puts the other half of the money into an account that grows continuously at an annual rate of 0.8%. Which function represents the value of Kelly-Ann's investments after t years?
(1) $f(t) = 10,000(1.9)^t + 10,000e^{0.8t}$
(2) $f(t) = 10,000(1.009)^t + 10,000e^{0.008t}$
(3) $f(t) = 10,000(1.075)^{12t} + 10,000e^{0.8t}$
(4) $f(t) = 10,000(1.00075)^{12t} + 10,000e^{0.008t}$

18 _____

19. Which graph represents a polynomial function that contains $x^2 + 2x + 1$ as a factor?

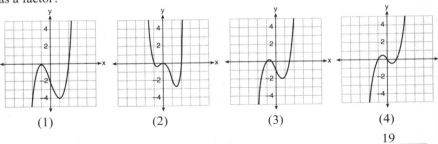

(1)　　　　　(2)　　　　　(3)　　　　　(4)

19 _____

20. Sodium iodide-131, used to treat certain medical conditions, has a half-life of 1.8 hours. The data table below shows the amount of sodium iodide-131, rounded to the nearest thousandth, as the dose fades over time.

Number of Half Lives	1	2	3	4	5
Amount of Sodium Iodide-131	139.000	69.500	34.750	17.375	8.688

What approximate amount of sodium iodide-131 will remain in the body after 18 hours?
(1) 0.001   (2) 0.136   (3) 0.271   (4) 0.543

20 _____

21. Which expression(s) are equivalent to $\dfrac{x^2 - 4x}{2x}$, where $x \neq 0$?

I. $\dfrac{x}{2} - 2$　　II. $\dfrac{x-4}{2}$　　III. $\dfrac{x-1}{2} - \dfrac{3}{2}$

(1) II, only   (2) I and II   (3) II and III   (4) I, II, and III

21 _____

22. Consider $f(x) = 4x^2 + 6x - 3$, and $p(x)$ defined by the graph. The difference between the values of the maximum of p and minimum of f is
(1) 0.25
(2) 1.25
(3) 3.25
(4) 10.25

22 _____

23. The scores on a mathematics college-entry exam are normally distributed with a mean of 68 and standard deviation 7.2. Students scoring higher than one standard deviation above the mean will not be enrolled in the mathematics tutoring program. How many of the 750 incoming students can be expected to be enrolled in the tutoring program?
(1) 631   (2) 512   (3) 238   (4) 119

23 _____

ALGEBRA 2 – COMMON CORE
January 2019

24. How many solutions exist for $\dfrac{1}{1-x^2} = -|3x-2|+5$?

(1) 1 (2) 2 (3) 3 (4) 4

24 _____

Part II

Answer all 8 questions in this part. Each correct answer will receive 2 credits. Clearly indicate the necessary steps, including appropriate formula substitutions, diagrams, graphs, charts, etc. Utilize the information provided for each question to determine your answer. Note that diagrams are not necessarily drawn to scale. For all questions in this part, a correct numerical answer with no work shown will receive only 1 credit. All answers should be written in pen, except for graphs and drawings, which should be done in pencil. [16]

25. Justify why $\dfrac{\sqrt[3]{x^2 y^5}}{\sqrt[4]{x^3 y^4}}$ is equivalent to $x^{\frac{-1}{12}} y^{\frac{2}{3}}$ using properties of rational exponents, where $x \neq 0$ and $y \neq 0$.

26. The zeros of a quartic polynomial function are 2, –2, 4, and –4. Use the zeros to construct a possible sketch of the function, on the set of axes below.

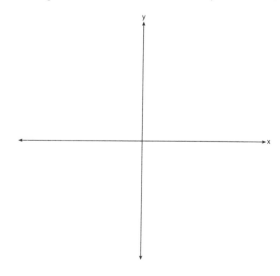

27. Erin and Christa were working on cubing binomials for math homework. Erin believed they could save time with a shortcut. She wrote down the rule below for Christa to follow.

$$(a + b)^3 = a^3 + b^3$$

Does Erin's shortcut always work? Justify your result algebraically.

28. The probability that a resident of a housing community opposes spending money for community improvement on plumbing issues is 0.8. The probability that a resident favors spending money on improving walkways given that the resident opposes spending money on plumbing issues is 0.85. Determine the probability that a randomly selected resident opposes spending money on plumbing issues and favors spending money on walkways.

29. Rowan is training to run in a race. He runs 15 miles in the first week, and each week following, he runs 3% more than the week before. Using a geometric series formula, find the total number of miles Rowan runs over the first ten weeks of training, rounded to the *nearest thousandth*.

30. The average monthly high temperature in Buffalo, in degrees Fahrenheit, can be modeled by the function $B(t) = 25.29\sin(0.4895t - 1.9752) + 55.2877$, where t is the month number (January = 1). State, to the *nearest tenth*, the average monthly rate of temperature change between August and November.

Explain its meaning in the given context.

31. Point $M\left(t, \frac{4}{7}\right)$ is located in the second quadrant on the unit circle. Determine the exact value of t.

32. On the grid below, graph the function $y = \log_2 (x - 3) + 1$

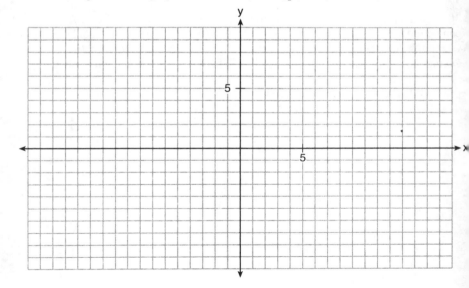

ALGEBRA 2 – COMMON CORE
January 2019
Part III

Answer all 4 questions in this part. Each correct answer will receive 4 credits. Clearly indicate the necessary steps, including appropriate formula substitutions, diagrams, graphs, charts, etc. Utilize the information provided for each question to determine your answer. Note that diagrams are not necessarily drawn to scale. For all questions in this part, a correct numerical answer with no work shown will receive only 1 credit. All answers should be written in pen, except for graphs and drawings, which should be done in pencil. [16]

33. Solve the following system of equations algebraically for all values of a, b, and c.

$$a + 4b + 6c = 23$$
$$a + 2b + c = 2$$
$$6b + 2c = a + 14$$

34. Given $a(x) = x^4 + 2x^3 + 4x - 10$ and $b(x) = x + 2$, determine $\dfrac{a(x)}{b(x)}$ in the form $q(x) + \dfrac{r(x)}{b(x)}$.

Is $b(x)$ a factor of $a(x)$? Explain.

35. A radio station claims to its advertisers that the mean number of minutes commuters listen to the station is 30. The station conducted a survey of 500 of their listeners who commute. The sample statistics are shown.

$\bar{x}$	29.11
s_x	20.718

A simulation was run 1000 times based upon the results of the survey. The results of the simulation appear below.

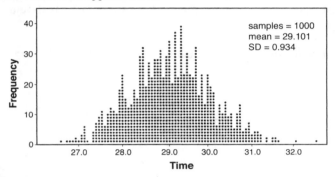

Based on the simulation results, is the claim that commuters listen to the station on average 30 minutes plausible? Explain your response including an interval containing the middle 95% of the data, rounded to the *nearest hundredth*.

36. Solve the given equation algebraically for all values of x.
$$3\sqrt{x} - 2x = -5$$

ALGEBRA 2 – COMMON CORE
January 2019
Part IV

Answer the question in this part. A correct answer will receive 6 credits. Clearly indicate the necessary steps, including appropriate formula substitutions, diagrams, graphs, charts, etc. Utilize the information provided to determine your answer. Note that diagrams are not necessarily drawn to scale. A correct numerical answer with no work shown will receive only 1 credit. All answers should be written in pen, except for graphs and drawings, which should be done in pencil. [6]

37. Tony is evaluating his retirement savings. He currently has $318,000 in his account, which earns an interest rate of 7% compounded annually. He wants to determine how much he will have in the account in the future, even if he makes no additional contributions to the account.

Write a function, $A(t)$, to represent the amount of money that will be in his account in t years.

Graph $A(t)$ where $0 \le t \le 20$ on the set of axes.

Tony's goal is to save $1,000,000. Determine algebraically, to the *nearest year*, how many years it will take for him to achieve his goal.

Explain how your graph of $A(t)$ confirms your answer.

ALGEBRA 2 – COMMON CORE
June 2019
Part I

Answer all 24 questions in this part. Each correct answer will receive 2 credits. No partial credit will be allowed. Utilize the information provided for each question to determine your answer. Note that diagrams are not necessarily drawn to scale. For each statement or question, choose the word or expression that, of those given, best completes the statement or answers the question. Record your answers in the space provided. [48]

1. A sociologist reviews randomly selected surveillance videos from a public park over a period of several years and records the amount of time people spent on a smartphone. The statistical procedure the sociologist used is called
(1) a census
(2) an experiment
(3) an observational study
(4) a sample survey

1 _____

2. Which statement(s) are true for all real numbers?
 I $(x-y)^2 = x^2 + y^2$
 II $(x+y)^3 = x^3 + 3xy + y^3$
(1) I, only (2) II, only (3) I and II (4) neither I nor II

2 _____

3. What is the solution set of the following system of equations?
$$y = 3x + 6$$
$$y = (x+4)^2 - 10$$
(1) $\{(-5, -9)\}$ (2) $\{(5, 21)\}$ (3) $\{(0, 6),(-5, -9)\}$ (4) $\{(0, 6),(5, 21)\}$

3 _____

4. Irma initially ran one mile in over ten minutes. She then began a training program to reduce her one-mile time. She recorded her one-mile time once a week for twelve consecutive weeks, as modeled in the accompanying graph.

Which statement regarding Irma's one-mile training program is correct?
(1) Her one-mile speed increased as the number of weeks increased.
(2) Her one-mile speed decreased as the number of weeks increased.
(3) If the trend continues, she will run under a six-minute mile by week thirteen.
(4) She reduced her one-mile time the most between weeks ten and twelve.

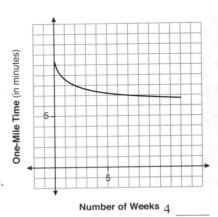

4 _____

5. A 7-year lease for office space states that the annual rent is $85,000 for the first year and will increase by 6% each additional year of the lease. What will the total rent expense be for the entire 7-year lease?
(1) $42,809.63 (2) $90,425.53 (3) $595,000.00 (4) $713,476.20

5 _____

ALGEBRA 2 – COMMON CORE
June 2019

6. The graph of $y = f(x)$ is shown. Which expression defines $f(x)$?

(1) $2x$ (3) $5\left(2^{\frac{x}{2}}\right)$
(2) $5(2^x)$ (4) $5(2^{2x})$

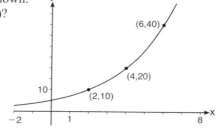

6 _____

7. Given $P(x) = x^3 - 3x^2 - 2x + 4$, which statement is true?
(1) $(x - 1)$ is a factor because $P(-1) = 2$.
(2) $(x + 1)$ is a factor because $P(-1) = 2$.
(3) $(x + 1)$ is a factor because $P(1) = 0$.
(4) $(x - 1)$ is a factor because $P(1) = 0$.

7 _____

8. For $x \geq 0$, which equation is *false*?

(1) $\left(x^{\frac{3}{2}}\right)^2 = \sqrt[4]{x^3}$

(2) $(x^3)^{\frac{1}{4}} = \sqrt[4]{x^3}$

(3) $\left(x^{\frac{3}{2}}\right)^{\frac{1}{2}} = \sqrt[4]{x^3}$

(4) $\left(x^{\frac{2}{3}}\right)^2 = \sqrt[3]{x^4}$

8 _____

9. What is the inverse of the function $y = 4x + 5$?

(1) $x = \frac{1}{4}y - \frac{5}{4}$

(2) $y = \frac{1}{4}x - \frac{5}{4}$

(3) $y = 4x - 5$

(4) $y = \frac{1}{4x + 5}$

9 _____

10. Which situation could be modeled using a geometric sequence?
(1) A cell phone company charges $30.00 per month for 2 gigabytes of data and $12.50 for each additional gigabyte of data.
(2) The temperature in your car is 79°. You lower the temperature of your air conditioning by 2° every 3 minutes in order to find a comfortable temperature.
(3) David's parents have set a limit of 50 minutes per week that he may play online games during the school year. However, they will increase his time by 5% per week for the next ten weeks.
(4) Sarah has $100.00 in her piggy bank and saves an additional $15.00 each week.

10 _____

11. The completely factored form of $n^4 - 9n^2 + 4n^3 - 36n - 12n^2 + 108$ is
(1) $(n^2 - 9)(n + 6)(n - 2)$
(2) $(n + 3)(n - 3)(n + 6)(n - 2)$
(3) $(n - 3)(n - 3)(n + 6)(n - 2)$
(4) $(n + 3)(n - 3)(n - 6)(n + 2)$

11 _____

12. What is the solution when the equation $wx^2 + w = 0$ is solved for x, where w is a positive integer?
(1) -1 (2) 0   (3) 6   (4) $\pm i$

12 _____

13. A group of students was trying to determine the proportion of candies in a bag that are blue. The company claims that 24% of candies in bags are blue. A simulation was run 100 times with a sample size of 50, based on the premise that 24% of the candies are blue. The approximately normal results of the simulation are shown in the dot plot below.

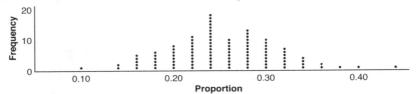

The simulation results in a mean of 0.254 and a standard deviation of 0.060. Based on this simulation, what is a plausible interval containing the middle 95% of the data?
(1) (0.194, 0.314)
(2) (0.134, 0.374)
(3) (−0.448, 0.568)
(4) (0.254, 0.374)   13 _____

14. Selected values for the functions f and g are shown in the accompanying tables.

A solution to the equation $f(x) = g(x)$ is
(1) 0
(2) 2.53
(3) 3.01
(4) 8.52

x	f(x)
−3.12	−4.88
0	−6
1.23	−4.77
8.52	2.53
9.01	3.01

x	g(x)
−2.01	−1.01
0	0.58
8.52	2.53
13.11	3.01
16.52	3.29

14 _____

15. The expression $6 - (3x - 2i)^2$ is equivalent to
(1) $-9x^2 + 12xi + 10$
(2) $9x^2 - 12xi + 2$
(3) $-9x^2 + 10$
(4) $-9x^2 + 12xi - 4i + 6$ 15 _____

16. A number, minus twenty times its reciprocal, equals eight. The number is
(1) 10 or −2   (2) 10 or 2   (3) −10 or −2   (4) −10 or 2   16 _____

17. A savings account, S, has an initial value of $50. The account grows at a 2% interest rate compounded n times per year, t, according to the function below.
$$S(t) = 50\left(1 + \frac{.02}{n}\right)^{nt}$$

Which statement about the account is correct?
(1) As the value of n increases, the amount of interest per year decreases.
(2) As the value of n increases, the value of the account approaches the function $S(t) = 50e^{0.02t}$.
(3) As the value of n decreases to one, the amount of interest per year increases.
(4) As the value of n decreases to one, the value of the account approaches the function $S(t) = 50(1 - 0.02)^t$.   17 _____

ALGEBRA 2 – COMMON CORE
June 2019

18. There are 400 students in the senior class at Oak Creek High School. All of these students took the SAT. The distribution of their SAT scores is approximately normal. The number of students who scored within 2 standard deviations of the mean is approximately
(1) 75 (2) 95 (3) 300 (4) 380   18 ____

19. The solution set for the equation $b = \sqrt{2b^2 - 64}$ is
(1) $\{-8\}$ (2) $\{8\}$ (3) $\{\pm 8\}$ (4) $\{\ \}$ 19 ____

20. Which table best represents an exponential relationship?

x	y
1	8
2	4
3	2
4	1
5	$\frac{1}{2}$

(1)

x	y
8	0
4	1
0	2
-4	3
-8	4

(2)

x	y
0	0
1	1
2	4
3	9
4	16

(3)

x	y
1	1
2	8
3	27
4	64
5	125

(4)

20 ____

21. A sketch of $r(x)$ is shown.

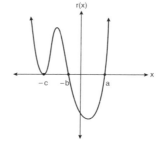

An equation for $r(x)$ could be
(1) $r(x) = (x - a)(x + b)(x + c)$
(2) $r(x) = (x + a)(x - b)(x - c)^2$
(3) $r(x) = (x + a)(x - b)(x - c)$
(4) $r(x) = (x - a)(x + b)(x + c)^2$

21 ____

22. The temperature, in degrees Fahrenheit, in Times Square during a day in August can be predicted by the function $T(x) = 8\sin(0.3x - 3) + 74$, where x is the number of hours after midnight. According to this model, the predicted temperature, to the *nearest degree* Fahrenheit, at 7 P.M. is
(1) 68 (2) 74 (3) 77 (4) 81   22 ____

23. Consider the system of equations:
$$x + y - z = 6$$
$$2x - 3y + 2z = -19$$
$$-x + 4y - z = 17$$
Which number is *not* the value of any variable in the solution of the system?
(1) –1 (2) 2 (3) 3 (4) –4   23 ____

24. Camryn puts $400 into a savings account that earns 6% annually. The amount in her account can be modeled by $C(t) = 400(1.06)^t$ where t is the time in years. Which expression best approximates the amount of money in her account using a weekly growth rate?
(1) $400(1.001153846)^t$
(2) $400(1.001121184)^t$
(3) $400(1.001153846)^{52t}$
(4) $400(1.001121184)^{52t}$

24 ____

ALGEBRA 2 – COMMON CORE
June 2019
Part II

Answer all 8 questions in this part. Each correct answer will receive 2 credits. Clearly indicate the necessary steps, including appropriate formula substitutions, diagrams, graphs, charts, etc. Utilize the information provided for each question to determine your answer. Note that diagrams are not necessarily drawn to scale. For all questions in this part, a correct numerical answer with no work shown will receive only 1 credit. All answers should be written in pen, except for graphs and drawings, which should be done in pencil. [16]

25. The table shows the number of hours of daylight on the first day of each month in Rochester, NY.

Given the data, what is the average rate of change in hours of daylight per month from January 1st to April 1st?

Month	Hours of Daylight
Jan.	9.4
Feb.	10.6
March	11.9
April	13.9
May	14.7
June	15.4
July	15.1
Aug.	13.9
Sept.	12.5
Oct.	11.1
Nov.	9.7
Dec.	9.0

Interpret what this means in the context of the problem.

26. Algebraically solve for x: $\dfrac{7}{2x} - \dfrac{2}{x+1} = \dfrac{1}{4}$

27. Graph $f(x) = \log_2(x+6)$ on the set of axes below.

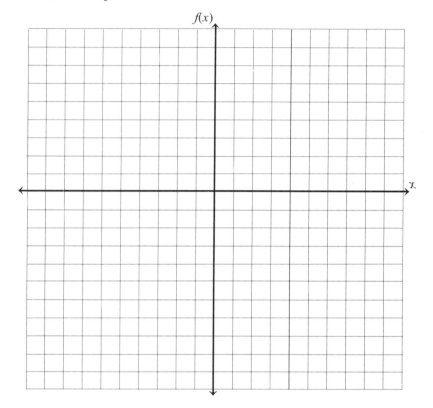

28. Given $\tan \theta = \dfrac{7}{24}$ and θ terminates in Quadrant III, determine the value of $\cos \theta$.

29. Kenzie believes that for $x \geq 0$, the expression $\left(\sqrt[7]{x^2}\right)\left(\sqrt[5]{x^3}\right)$ is equivalent to $\sqrt[35]{x^6}$. Is she correct? Justify your response algebraically.

30. When the function $p(x)$ is divided by $x - 1$ the quotient is $x^2 + 7 + \dfrac{5}{x-1}$. State $p(x)$ in standard form.

31. Write a recursive formula for the sequence 6, 9, 13.5, 20.25, . . .

ALGEBRA 2 – COMMON CORE
June 2019

32. Robin flips a coin 100 times. It lands heads up 43 times, and she wonders if the coin is unfair. She runs a computer simulation of 750 samples of 100 fair coin flips. The output of the proportion of heads is shown.

Do the results of the simulation provide strong evidence that Robin's coin is unfair? Explain your answer.

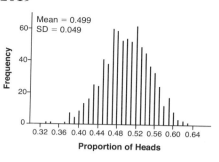

Part III

Answer all 4 questions in this part. Each correct answer will receive 4 credits. Clearly indicate the necessary steps, including appropriate formula substitutions, diagrams, graphs, charts, etc. Utilize the information provided for each question to determine your answer. Note that diagrams are not necessarily drawn to scale. For all questions in this part, a correct numerical answer with no work shown will receive only 1 credit. All answers should be written in pen, except for graphs and drawings, which should be done in pencil. [16]

33. Factor completely over the set of integers: $16x^4 - 81$

Sara graphed the polynomial $y = 16x^4 - 81$ and stated "All the roots of $y = 16x^4 - 81$ are real." Is Sara correct? Explain your reasoning.

34. The half-life of a radioactive substance is 15 years.

Write an equation that can be used to determine the amount, $s(t)$, of 200 grams of this substance that remains after t years.

Determine algebraically, to the *nearest year*, how long it will take for $\frac{1}{10}$ of this substance to remain.

35. Determine an equation for the parabola with focus $(4, -1)$ and directrix $y = -5$. (Use of the grid below is optional.)

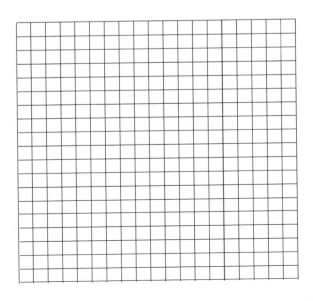

ALGEBRA 2 – COMMON CORE
June 2019

36. Juan and Filipe practice at the driving range before playing golf. The number of wins and corresponding practice times for each player are shown in the table below.

	Juan Wins	Filipe Wins
Short Practice Time	8	10
Long Practice Time	15	12

Given that the practice time was long, determine the exact probability that Filipe wins the next match.

Determine whether or not the two events "Filipe wins" and "long practice time" are independent. Justify your answer.

Part IV
Answer the question in this part. A correct answer will receive 6 credits. Clearly indicate the necessary steps, including appropriate formula substitutions, diagrams, graphs, charts, etc. Utilize the information provided to determine your answer. Note that diagrams are not necessarily drawn to scale. A correct numerical answer with no work shown will receive only 1 credit. All answers should be written in pen, except for graphs and drawings, which should be done in pencil. [6]

37. Griffin is riding his bike down the street in Churchville, N.Y. at a constant speed, when a nail gets caught in one of his tires. The height of the nail above the ground, in inches, can be represented by the trigonometric function $f(t) = -13\cos(0.8\pi t) + 13$, where t represents the time (in seconds) since the nail first became caught in the tire.

Determine the period of $f(t)$.

Interpret what the period represents in this context.

Question 37 is continued on the next page.

Question 37 continued

On the grid below, graph *at least one* cycle of f(t) that includes the y-intercept of the function.

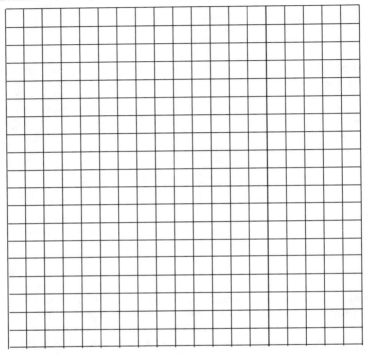

Does the height of the nail ever reach 30 inches above the ground? Justify your answer.

ALGEBRA 2 – COMMON CORE
August 2019
Part I

Answer all 24 questions in this part. Each correct answer will receive 2 credits. No partial credit will be allowed. Utilize the information provided for each question to determine your answer. Note that diagrams are not necessarily drawn to scale. For each statement or question, choose the word or expression that, of those given, best completes the statement or answers the question. Record your answers in the space provided. [48]

1. When the expression $(x + 2)^2 + 4(x + 2) + 3$ is rewritten as the product of two binomials, the result is
(1) $(x + 3)(x + 1)$
(2) $(x + 5)(x + 3)$
(3) $(x + 2)(x + 2)$
(4) $(x + 6)(x + 1)$

2. The first term of a geometric sequence is 8 and the fourth term is 216. What is the sum of the first 12 terms of the corresponding series?
(1) 236,192
(2) 708,584
(3) 2,125,760
(4) 6,377,288

3. Perry invested in property that cost him $1500. Five years later it was worth $3000, and 10 years from his original purchase, it was worth $6000. Assuming the growth rate remains the same, which type of function could he create to find the value of his investment 30 years from his original purchase?
(1) exponential function
(2) linear function
(3) quadratic function
(4) trigonometric function

4. If $(a^3 + 27) = (a + 3)(a^2 + ma + 9)$, then m equals
(1) –9
(2) –3
(3) 3
(4) 6

5. If $\cos \theta = -\frac{3}{4}$ and θ is in Quadrant III, then $\sin \theta$ is equivalent to
(1) $-\frac{\sqrt{7}}{4}$
(2) $\frac{\sqrt{7}}{4}$
(3) $-\frac{5}{4}$
(4) $\frac{5}{4}$

6. A veterinary pharmaceutical company plans to test a new drug to treat a common intestinal infection among puppies. The puppies are randomly assigned to two equal groups. Half of the puppies will receive the drug, and the other half will receive a placebo. The veterinarians monitor the puppies. This is an example of which study method?
(1) census
(2) observational study
(3) survey
(4) controlled experiment

7. The expression $2 - \frac{x-1}{x+2}$ is equivalent to
(1) $1 - \frac{3}{x+2}$
(2) $1 + \frac{3}{x+2}$
(3) $1 - \frac{1}{x+2}$
(4) $1 + \frac{1}{x+2}$

8. Which description could represent the graph of $f(x) = 4x^2(x + a) - x - a$, if a is an integer?
(1) As $x \to -\infty$, $f(x) \to \infty$, as $x \to \infty$, $f(x) \to \infty$, and the graph has 3 x-intercepts.
(2) As $x \to -\infty$, $f(x) \to -\infty$, as $x \to \infty$, $f(x) \to \infty$, and the graph has 3 x-intercepts.
(3) As $x \to -\infty$, $f(x) \to \infty$, as $x \to \infty$, $f(x) \to -\infty$, and the graph has 4 x-intercepts.
(4) As $x \to -\infty$, $f(x) \to -\infty$, as $x \to \infty$, $f(x) \to \infty$, and the graph has 4 x-intercepts.

8 _____

9. After Roger's surgery, his doctor administered pain medication in the following amounts in milligrams over four days.

Day (n)	1	2	3	4
Dosage (m)	2000	1680	1411.2	1185.4

How can this sequence best be modeled recursively?
(1) $m_1 = 2000$
  $m_n = m_{n-1} - 320$
(2) $m_n = 2000(0.84)^{n-1}$
(3) $m_1 = 2000$
  $m_n = (0.84)m_{n-1}$
(4) $m_n = 2000(0.84)^{n+1}$

9 _____

10. The expression $\dfrac{9x^2 - 2}{3x + 1}$ is equivalent to
(1) $3x - 1 - \dfrac{1}{3x+1}$
(2) $3x - 1 + \dfrac{1}{3x+1}$
(3) $3x + 1 - \dfrac{1}{3x+1}$
(4) $3x + 1 + \dfrac{1}{3x+1}$

10 _____

11. If $f(x)$ is an even function, which function must also be even?
(1) $f(x - 2)$ (2) $f(x) + 3$ (3) $f(x + 1)$ (4) $f(x + 1) + 3$

11 _____

12. The average monthly temperature of a city can be modeled by a cosine graph. Melissa has been living in Phoenix, Arizona, where the average annual temperature is 75°F. She would like to move, and live in a location where the average annual temperature is 62°F. When examining the graphs of the average monthly temperatures for various locations, Melissa should focus on the
(1) amplitude (2) horizontal shift (3) period (4) midline

12 _____

13. Consider the probability statements regarding events A and B below.
$P(A \text{ or } B) = 0.3$;
$P(A \text{ and } B) = 0.2$, and
$P(A \mid B) = 0.8$

What is $P(B)$?
(1) 0.1 (2) 0.25 (3) 0.375 (4) 0.667

13 _____

14. Given $y > 0$, the expression $\sqrt{3x^2 y} \cdot \sqrt[3]{27x^3 y^2}$ is equivalent to
(1) $81x^5 y^3$ (2) $3^{1.5} x^2 y$ (3) $3^{\frac{5}{2}} x^2 y^{\frac{5}{3}}$ (4) $3^{\frac{3}{2}} x^2 y^{\frac{7}{6}}$

14 _____

ALGEBRA 2 – COMMON CORE
August 2019

15. What is the solution set of the equation $\dfrac{10}{x^2 - 2x} + \dfrac{4}{x} = \dfrac{5}{x-2}$?
(1) {0, 2}   (2) {0}   (3) {2}   (4) { }   15 _____

16. What are the solution(s) to the system of equations shown below?
$$x^2 + y^2 = 5$$
$$y = 2x$$
(1) $x = 1$ and $x = -1$ (3) (1, 2) and (−1, −2)
(2) $x = 1$ (4) (1, 2), only   16 _____

17. If $5000 is put into a savings account that pays 3.5% interest compounded monthly, how much money, to the *nearest ten cents*, would be in that account after 6 years, assuming no money was added or withdrawn?
(1) $5177.80   (2) $5941.30   (3) $6146.30   (4) $6166.50   17 _____

18. The Fahrenheit temperature, $F(t)$, of a heated object at time t, in minutes, can be modeled by the function below. F_s is the surrounding temperature, F_0 is the initial temperature of the object, and k is a constant.
$$F(t) = F_s + (F_0 - F_s)e^{-kt}$$
Coffee at a temperature of 195°F is poured into a container. The room temperature is kept at a constant 68°F and $k = 0.05$. Coffee is safe to drink when its temperature is, at most, 120°F. To the *nearest minute*, how long will it take until the coffee is safe to drink?
(1) 7   (2) 10   (3) 11   (4) 18   18 _____

19. The mean intelligence quotient (IQ) score is 100, with a standard deviation of 15, and the scores are normally distributed. Given this information, the approximate percentage of the population with an IQ greater than 130 is closest to
(1) 2%   (2) 31%   (3) 48%   (4) 95%   19 _____

20. After examining the functions $f(x) = \ln(x + 2)$ and $g(x) = e^{x-1}$ over the interval (−2, 3], Lexi determined that the correct number of solutions to the equation $f(x) = g(x)$ is
(1) 1   (2) 2   (3) 3   (4) 0   20 _____

21. Evan graphed a cubic function, $f(x) = ax^3 + bx^2 + cx + d$, and determined the roots of $f(x)$ to be ±1 and 2. What is the value of b, if $a = 1$?
(1) 1   (2) 2   (3) −1   (4) −2   21 _____

22. The equation $t = \dfrac{1}{0.0105} \ln\left(\dfrac{A}{5000}\right)$ relates time, t, in years, to the amount of money, A, earned by a $5000 investment. Which statement accurately describes the relationship between the average rates of change of t on the intervals [6000, 8000] and [9000, 12,000]?
(1) A comparison cannot be made because the intervals are different sizes.
(2) The average rate of change is equal for both intervals.
(3) The average rate of change is larger for the interval [6000, 8000].
(4) The average rate of change is larger for the interval [9000, 12,000].   22 _____

23. What is the inverse of $f(x) = \dfrac{x}{x+2}$, where $x \neq -2$?

(1) $f^{-1}(x) = \dfrac{2x}{x-1}$

(2) $f^{-1}(x) = \dfrac{-2x}{x-1}$

(3) $f^{-1}(x) = \dfrac{x}{x-2}$

(4) $f^{-1}(x) = \dfrac{-x}{x-2}$

23 _____

24. A study of black bears in the Adirondacks reveals that their population can be represented by the function $P(t) = 3500(1.025)^t$, where t is the number of years since the study began. Which function is correctly rewritten to reveal the monthly growth rate of the black bear population?

(1) $P(t) = 3500(1.00206)^{12t}$

(2) $P(t) = 3500(1.00206)^{\frac{t}{12}}$

(3) $P(t) = 3500(1.34489)^{12t}$

(4) $P(t) = 3500(1.34489)^{\frac{t}{12}}$

24 _____

Part II

Answer all 8 questions in this part. Each correct answer will receive 2 credits. Clearly indicate the necessary steps, including appropriate formula substitutions, diagrams, graphs, charts, etc. Utilize the information provided for each question to determine your answer. Note that diagrams are not necessarily drawn to scale. For all questions in this part, a correct numerical answer with no work shown will receive only 1 credit. All answers should be written in pen, except for graphs and drawings, which should be done in pencil. [16]

25. At Andrew Jackson High School, students are only allowed to enroll in AP U.S. History if they have already taken AP World History or AP European History. Out of 825 incoming seniors, 165 took AP World History, 66 took AP European History, and 33 took both. Given this information, determine the probability a randomly selected incoming senior is allowed to enroll in AP U.S. History.

26. Explain what a rational exponent, such as $\frac{5}{2}$ means. Use this explanation to evaluate $9^{\frac{5}{2}}$.

27. Write $-\frac{1}{2}i^3(\sqrt{-9} - 4) - 3i^2$ in simplest $a + bi$ form.

28. A person's lung capacity can be modeled by the function $C(t) = 250\sin\left(\frac{2\pi}{5}t\right) + 2450$, where $C(t)$ represents the volume in mL present in the lungs after t seconds. State the maximum value of this function over one full cycle, and explain what this value represents.

ALGEBRA 2 – COMMON CORE
August 2019

29. Determine for which polynomial(s) $(x + 2)$ is a factor. Explain your answer.

$P(x) = x^4 - 3x^3 - 16x - 12$
$Q(x) = x^3 - 3x^2 - 16x - 12$

30. On July 21, 2016, the water level in Puget Sound, WA reached a high of 10.1 ft at 6 a.m. and a low of –2 ft at 12:30 p.m. Across the country in Long Island, NY, Shinnecock Bay's water level reached a high of 2.5 ft at 10:42 p.m. and a low of –0.1 ft at 5:31 a.m.

The water levels of both locations are affected by the tides and can be modeled by sinusoidal functions. Determine the difference in amplitudes, in feet, for these two locations.

31. Write a recursive formula, a_n, to describe the sequence graphed.

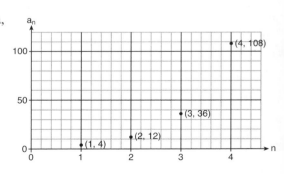

ALGEBRA 2 – COMMON CORE
August 2019

32. Sketch the graphs of $r(x) = \dfrac{1}{x}$ and $a(x) = |x| - 3$ on the set of axes below. Determine, to the *nearest tenth*, the positive solution of $r(x) = a(x)$.

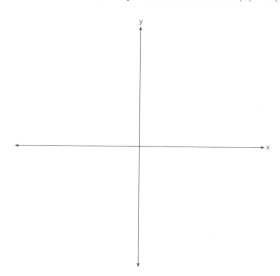

Part III

Answer all 4 questions in this part. Each correct answer will receive 4 credits. Clearly indicate the necessary steps, including appropriate formula substitutions, diagrams, graphs, charts, etc. Utilize the information provided for each question to determine your answer. Note that diagrams are not necessarily drawn to scale. For all questions in this part, a correct numerical answer with no work shown will receive only 1 credit. All answers should be written in pen, except for graphs and drawings, which should be done in pencil. [16]

33. A population of 950 bacteria grows continuously at a rate of 4.75% per day.

Write an exponential function, $N(t)$, that represents the bacterial population after t days and explain the reason for your choice of base.

Determine the bacterial population after 36 hours, to the *nearest bacterium*.

34. Write an equation for a sine function with an amplitude of 2 and a period of $\frac{\pi}{2}$.

On the grid below, sketch the graph of the equation in the interval 0 to 2π.

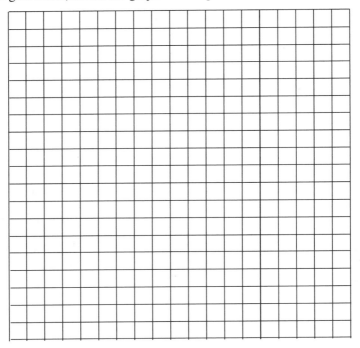

35. Mary bought a pack of candy. The manufacturer claims that 30% of the candies manufactured are red. In her pack, 14 of the 60 candies are red. She ran a simulation of 300 samples, assuming the manufacturer is correct. The results are shown below.

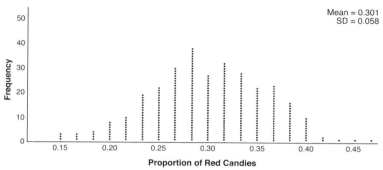

Based on the simulation, determine the middle 95% of plausible values that the proportion of red candies in a pack is within.

Based on the simulation, is it unusual that Mary's pack had 14 red candies out of a total of 60? Explain

36. a) Algebraically determine the roots, in simplest $a + bi$ form, to the equation below.
$$x^2 - 2x + 7 = 4x - 10$$

b) Consider the system of equations
$$y = x^2 - 2x + 7$$
$$y = 4x - 10$$
The graph of this system confirms the solution from part a is imaginary. Explain why.

ALGEBRA 2 – COMMON CORE
August 2019
Part IV

Answer the question in this part. A correct answer will receive 6 credits. Clearly indicate the necessary steps, including appropriate formula substitutions, diagrams, graphs, charts, etc. Utilize the information provided to determine your answer. Note that diagrams are not necessarily drawn to scale. A correct numerical answer with no work shown will receive only 1 credit. All answers should be written in pen, except for graphs and drawings, which should be done in pencil. [6]

Beaufort Wind Scale

Beaufort Number	Force of Wind
0	Calm
1	Light air
2	Light breeze
3	Gentle breeze
4	Moderate breeze
5	Fresh breeze
6	Steady breeze
7	Moderate gale
8	Fresh gale
9	Strong gale
10	Whole gale
11	Storm
12	Hurricane

37. The Beaufort Wind Scale was devised by British Rear Admiral Sir Francis Beaufort, in 1805 based upon observations of the effects of the wind. Beaufort numbers, B, are determined by the equation $B = 1.69\sqrt{s + 4.45} - 3.49$, where s is the speed of the wind in mph, and B is rounded to the nearest integer from 0 to 12.

Using the table, classify the force of wind at a speed of 30 mph. Justify your answer.

In 1946, the scale was extended to accommodate strong hurricanes. A strong hurricane received a B value of exactly 15. Algebraically determine the value of s, to the *nearest mph*.

Any B values that round to 10 receive a Beaufort number of 10. Using technology, find an approximate range of wind speeds, to the *nearest mph*, associated with a Beaufort number of 10.

ALGEBRA 2 – COMMON CORE
January 2020
Part I

Answer all 24 questions in this part. Each correct answer will receive 2 credits. No partial credit will be allowed. Utilize the information provided for each question to determine your answer. Note that diagrams are not necessarily drawn to scale. For each statement or question, choose the word or expression that, of those given, best completes the statement or answers the question. Record your answers in the space provided. [48]

1. The expression $\sqrt[4]{81x^8y^6}$ is equivalent to
(1) $3x^2y^{\frac{3}{2}}$ (2) $3x^4y^2$ (3) $9x^2y^{\frac{3}{2}}$ (4) $9x^4y^2$ 1 _____

2. Chet has $1200 invested in a bank account modeled by the function $P(n) = 1200(1.002)^n$, where $P(n)$ is the value of his account, in dollars, after n months. Chet's debt is modeled by the function $Q(n) = 100n$, where $Q(n)$ is the value of debt, in dollars, after n months. After n months, which function represents Chet's net worth, $R(n)$?
(1) $R(n) = 1200(1.002)^n + 100n$ (3) $R(n) = 1200(1.002)^n - 100n$
(2) $R(n) = 1200(1.002)^{12n} + 100n$ (4) $R(n) = 1200(1.002)^{12n} - 100n$ 2 _____

3. Emmeline is working on one side of a polynomial identity proof used to form Pythagorean triples. Her work is shown below:
$$(5x)^2 + (5x^2 - 5)^2$$
Step 1: $25x^2 + (5x^2 - 5)^2$ Step 3: $50x^2 + 25$
Step 2: $25x^2 + 25x^2 + 25$ Step 4: $75x^2$

What statement is true regarding Emmeline's work?
(1) Emmeline's work is entirely correct.
(2) There is a mistake in step 2, only.
(3) There are mistakes in step 2 and step 4.
(4) There is a mistake in step 4, only.   3 _____

4. Susan won $2,000 and invested it into an account with an annual interest rate of 3.2%. If her investment were compounded monthly, which expression best represents the value of her investment after t years?
(1) $2000(1.003)^{12t}$ (2) $2000(1.032)^{\frac{t}{12}}$ (3) $2064^{\frac{t}{12}}$ (4) $\frac{2000(1.032)^t}{12}$ 4 _____

5. Consider the end behavior description.
• as $x \to -\infty$, $f(x) \to \infty$
• as $x \to \infty$, $f(x) \to -\infty$
Which function satisfies the given conditions?

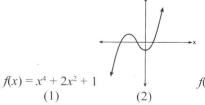
$f(x) = x^4 + 2x^2 + 1$
(1)   (2)

$f(x) = -x^3 + 2x - 6$
(3)   (4)

5 _____

ALGEBRA 2 – COMMON CORE
January 2020

6. The expression $(x + a)^2 + 5(x + a) + 4$ is equivalent to
(1) $(a + 1)(a + 4)$
(2) $(x + 1)(x + 4)$
(3) $(x + a + 1)(x + a + 4)$
(4) $x^2 + a^2 + 5x + 5a + 4$

6 _____

7. Given $x \neq -2$, the expression $\dfrac{2x^2 + 5x + 8}{x + 2}$ is equivalent to
(1) $2x^2 + \dfrac{9}{x + 2}$
(2) $2x + \dfrac{7}{x + 2}$
(3) $2x + 1 + \dfrac{6}{x + 2}$
(4) $2x + 9 - \dfrac{10}{x + 2}$

7 _____

8. Which situation best describes conditional probability?
(1) finding the probability of an event occurring two or more times
(2) finding the probability of an event occurring only once
(3) finding the probability of two independent events occurring at the same time
(4) finding the probability of an event occurring given another event had already occurred

8 _____

9. Which expression is *not* a solution to the equation $2^t = \sqrt{10}$?
(1) $\dfrac{1}{2}\log_2 10$
(2) $\log_2 \sqrt{10}$
(3) $\log_4 10$
(4) $\log_{10} 4$

9 _____

10. What is the solution set of $x = \sqrt{3x + 40}$?
(1) $\{-5, 8\}$
(2) $\{8\}$
(3) $\{-4, 10\}$
(4) $\{\ \}$

10 _____

11. Consider the data in the table below.

	Right Handed	Left Handed
Male	87	13
Female	89	11

What is the probability that a randomly selected person is male given the person is left handed?
(1) $\dfrac{13}{200}$
(2) $\dfrac{13}{100}$
(3) $\dfrac{13}{50}$
(4) $\dfrac{13}{24}$

11 _____

12. The function $N(x) = 90(0.86)^x + 69$ can be used to predict the temperature of a cup of hot chocolate in degrees Fahrenheit after x minutes. What is the approximate average rate of change of the temperature of the hot chocolate, in degrees per minute, over the interval $[0, 6]$?
(1) -8.93
(2) -0.11
(3) 0.11
(4) 8.93

12 _____

13. A recursive formula for the sequence 40, 30, 22.5, ... is
(1) $g_n = 40\left(\dfrac{3}{4}\right)^n$
(1) $g_n = 40\left(\dfrac{3}{4}\right)^{n-1}$
(2) $g_1 = 40$
  $g_n = g_{n-1} - 10$
(4) $g_1 = 40$
  $g_n = \dfrac{3}{4} g_{n-1}$

13 _____

14. The J & B candy company claims that 45% of the candies it produces are blue, 30% are brown, and 25% are yellow. Each bag holds 65 candies. A simulation was run 200 times, each of sample size 65, based on the premise that 45% of the candies are blue. The results of the simulation are shown below.

Bonnie purchased a bag of J & B's candy and counted 24 blue candies. What inference can be made regarding a bag of J & B's with only 24 blue candies?

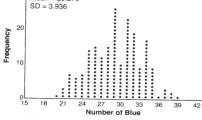

(1) The company is not meeting their production standard.
(2) Bonnie's bag was a rarity and the company should not be concerned.
(3) The company should change their claim to 37% blue candies are produced.
(4) Bonnie's bag is within the middle 95% of the simulated data supporting the company's claim.

14 _____

15. Which investigation technique is most often used to determine if a single variable has an impact on a given population?
(1) observational study   (3) controlled experiment
(2) random survey   (4) formal interview

15 _____

16. As θ increases from $-\frac{\pi}{2}$ to 0 radians, the value of cos θ will
(1) decrease from 1 to 0   (3) increase from −1 to 0
(2) decrease from 0 to −1   (4) increase from 0 to 1

16 _____

17. Consider the following patterns:
    I. 16, −12, 9, −6.75, ...   III. 6, 18, 30, 42, ...
    II. 1, 4, 9, 16, ...   IV. $\frac{1}{2}, \frac{2}{3}, \frac{3}{4}, \frac{4}{5}, ...$

Which pattern is geometric?
(1) I   (2) II   (3) III   (4) IV

17 _____

18. Consider the accompanying system.
Which value is *not* in the solution, (x, y, z), of the system?

$x + y + z = 9$
$x - y - z = -1$
$x - y + z = 21$

(1) −8   (2) −6   (3) 11   (4) 4

18 _____

19. Which statement regarding polynomials and their zeros is true?
(1) $f(x) = (x^2 − 1)(x + a)$ has zeros of 1 and −a, only.
(2) $f(x) = x^3 − ax^2 + 16x − 16a$ has zeros of 4 and a, only.
(3) $f(x) = (x^2 + 25)(x + a)$ has zeros of ±5 and −a.
(4) $f(x) = x^3 − ax^2 − 9x + 9a$ has zeros of ±3 and a.

19 _____

20. If a solution of $2(2x − 1) = 5x^2$ is expressed in simplest $a + bi$ form, the value of b is

(1) $\frac{\sqrt{6}}{5}i$ (2) $\frac{\sqrt{6}}{5}$ (3) $\frac{1}{5}i$ (4) $\frac{1}{5}$

20 _____

21. Which value, to the *nearest tenth*, is the *smallest* solution of $f(x) = g(x)$ if $f(x) = 3\sin\left(\frac{1}{2}x\right) - 1$ and $g(x) = x^3 - 2x + 1$?
(1) −3.6   (2) −2.1   (3) −1.8   (4) 1.4   21 _____

22. Expressed in simplest $a + bi$ form, $(7 - 3i) + (x - 2i)^2 - (4i + 2x^2)$ is
(1) $(3 - x^2) - (4x + 7)i$ (3) $(3 - x^2) - 7i$
(2) $(3 + 3x^2) - (4x + 7)i$ (4) $(3 + 3x^2) - 7i$ 22 _____

23. Written in simplest form, the fraction $\dfrac{x^3 - 9x}{9 - x^2}$, where $x \neq \pm 3$, is equivalent to
(1) $-x$ (2) x (3) $\dfrac{-x(x+3)}{(3+x)}$ (4) $\dfrac{x(x-3)}{(3-x)}$ 23 _____

24. According to a study, 45% of Americans have type O blood. If a random number generator produces three-digit values from 000 to 999, which values would represent those having type O blood?
(1) between 000 and 045, inclusive
(2) between 000 and 444, inclusive
(3) between 000 and 449, inclusive
(4) between 000 and 450, inclusive   24 _____

Part II

Answer all 8 questions in this part. Each correct answer will receive 2 credits. Clearly indicate the necessary steps, including appropriate formula substitutions, diagrams, graphs, charts, etc. Utilize the information provided for each question to determine your answer. Note that diagrams are not necessarily drawn to scale. For all questions in this part, a correct numerical answer with no work shown will receive only 1 credit. All answers should be written in pen, except for graphs and drawings, which should be done in pencil. [16]

25. For n and $p > 0$, is the expression $\left(p^2 n^{\frac{1}{2}}\right)^8 \sqrt{p^5 n^4}$ equivalent to $p^{18} n^6 \sqrt{p}$? Justify your answer.

ALGEBRA 2 – COMMON CORE
January 2020

26. Show why $x - 3$ is a factor of $m(x) = x^3 - x^2 - 5x - 3$. Justify your answer.

27. Describe the transformation applied to the graph of $p(x) = 2^x$ that forms the new function $q(x) = 2^{x-3} + 4$.

28. The parabola $y = -\frac{1}{20}(x - 3)^2 + 6$ has its focus at $(3, 1)$. Determine and state the equation of the directrix. (The use of the grid below is optional.)

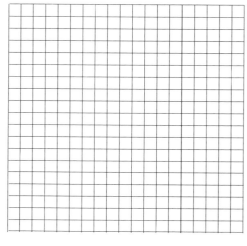

29. Given the geometric series $300 + 360 + 432 + 518.4 + \ldots$, write a geometric series formula, S_n, for the sum of the first n terms. Use the formula to find the sum of the first 10 terms, to the *nearest tenth*.

30. Visible light can be represented by sinusoidal waves. Three visible light waves are shown in the graph below. The midline of each wave is labeled ℓ.

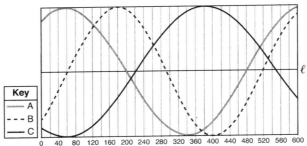

Based on the graph, which light wave has the longest period? Justify your answer.

31. Biologists are studying a new bacterium. They create a culture with 100 of the bacteria and anticipate that the number of bacteria will double every 30 hours. Write an equation for the number of bacteria, B, in terms of the number of hours, t, since the experiment began.

32. Graph $y = x^3 - 4x^2 + 2x + 7$ on the set of axes below.

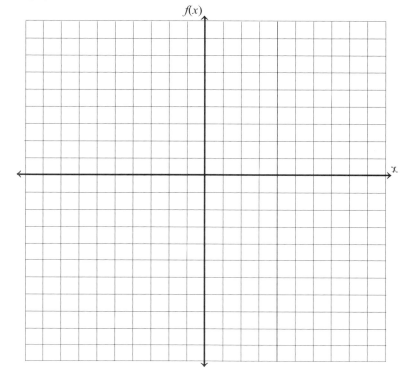

ALGEBRA 2 – COMMON CORE
January 2020
Part III

Answer all 4 questions in this part. Each correct answer will receive 4 credits. Clearly indicate the necessary steps, including appropriate formula substitutions, diagrams, graphs, charts, etc. Utilize the information provided for each question to determine your answer. Note that diagrams are not necessarily drawn to scale. For all questions in this part, a correct numerical answer with no work shown will receive only 1 credit. All answers should be written in pen, except for graphs and drawings, which should be done in pencil. [16]

33. Sonja is cutting wire to construct a mobile. She cuts 100 inches for the first piece, 80 inches for the second piece, and 64 inches for the third piece. Assuming this pattern continues, write an explicit equation for a_n, the length in inches of the nth piece.

Sonja only has 40 feet of wire to use for the project and wants to cut 20 pieces total for the mobile using her pattern. Will she have enough wire? Justify your answer.

34. Graph the following function on the axes below.

$f(x) = \log_3(2 - x)$

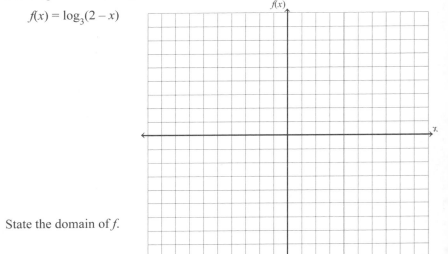

State the domain of f.

State the equation of the asymptote.

35. Algebraically solve the following system of equations.
$$(x-2)^2 + (y-3)^2 = 16$$
$$x + y - 1 = 0$$

36. The table below gives air pressures in kPa at selected altitudes above sea level measured in kilometers.

x	Altitude (km)	0	1	2	3	4	5
y	Air Pressure (kPa)	101	90	79	70	62	54

Write an exponential regression equation that models these data rounding all values to the *nearest thousandth*.

Use this equation to algebraically determine the altitude, to the *nearest hundredth* of a kilometer, when the air pressure is 29 kPa.

ALGEBRA 2 – COMMON CORE
January 2020
Part IV

Answer the question in this part. A correct answer will receive 6 credits. Clearly indicate the necessary steps, including appropriate formula substitutions, diagrams, graphs, charts, etc. Utilize the information provided to determine your answer. Note that diagrams are not necessarily drawn to scale. A correct numerical answer with no work shown will receive only 1 credit. All answers should be written in pen, except for graphs and drawings, which should be done in pencil. [6]

37. Sarah is fighting a sinus infection. Her doctor prescribed a nasal spray and an antibiotic to fight the infection. The active ingredients, in milligrams, remaining in the bloodstream from the nasal spray, $n(t)$, and the antibiotic, $a(t)$, are modeled in the functions below, where t is the time in hours since the medications were taken.

Determine which drug is made with a greater initial amount of active ingredient. Justify your answer.

$$n(t) = \frac{t+1}{t+5} + \frac{18}{t^2 + 8t + 15}$$

$$a(t) = \frac{9}{t+3}$$

Sarah's doctor told her to take both drugs at the same time. Determine algebraically the number of hours after taking the medications when both medications will have the same amount of active ingredient remaining in her bloodstream.